INVESTIGATING PLANTS

Hands-On, Low-Cost Laboratory Exercises in Plant Science

By Thomas R. Sinclair and Marty Johnson

Published by the National Association of Biology Teachers (NABT), 11250 Roger Bacon Drive, #19, Reston, Virginia 20190-5202, USA.

ISBN 0-941212-21-1.

Printed in the United States of America by Cadmus Journal Services.

Cover photographs courtesy of Biology Teacher Ed Reschke, Muskegon Community College, Muskegon, Michigan.

Acknowledgments

The exercises in this manual were developed with the enthusiastic participation of the seventh graders of Westwood Middle School, Gainesville, Florida. In the development of the exercises, these young people were quick to demonstrate those approaches that were successful and those that were not. Their exuberance in doing these activities was important in confirming that "Science is Fun."

Janas E. Sinclair made a significant contribution to this manual by skillfully and accurately preparing the illustrations that appear in this manual.

The encouragement and word processing skills of J. Annette Prasse during the development of the exercises and the preparation of the manual are gratefully acknowledged.

Patrons

The preparation and distribution of this manual was supported to an important extent by the:

Agricultural Research Service
U.S. Department of Agriculture
14th & Independence Ave., SW
Washington, DC 20250
Web Site: www.ars.usda.gov

Support for the publication of the manual was provided by:

AG in the Classroom
U.S. Department of Agriculture
4307 South Building
Washington, DC 20250-2200

AG in the Classroom can be contacted to obtain additional materials in support of classroom activities associated with biology and agriculture.

Financial contributions for the publication of this teachers' manual were made by two professional societies. The membership of these societies can also provide important resources in resolving questions concerning these exercises. Please contact either society for names of members within your area who can provide supplementary information.

Agronomic Science Foundation
677 South Segoe Road
Madison, Wisconsin 53711
Web Site: hhtp://www.agronomy.org

American Society of Plant Physiologists
15501 Monona Drive
Rockville, Maryland 20855-2768
Web Site: hhtp://aspp.org

About the Authors

Dr. Thomas R. Sinclair, Ph.D., was raised on a farm near Indianapolis, Indiana. He received his B.S. in Agricultural Science and M.S. in Plant Physiology from Purdue University, and his PhD in Field Crop Science from Cornell University. Dr. Sinclair is employed by the Agricultural Research Service, U.S. Department of Agriculture, and is an Adjunct Professor of the University of Florida in Gainesville, Florida.

Dr. Sinclair has been involved in plant research for more than 30 years investigating the interactions between crop plants and the environment. He has pursued his research worldwide with collaborative research in The Netherlands, Australia, Japan, Argentina, Italy, and Israel. He recently received an Honorary Doctorate Degree from the University of Padova, Italy.

Dr. Sinclair is married to a former middle school English teacher who offered useful advice and suggestions in developing these exercises. They have three children who were influential in stimulating Dr. Sinclair's interest in developing these hands-on exercises.

Marty Johnson, a native of El Paso, Texas, attended the University of Texas at both El Paso and Austin, receiving a B.A. degree in Biological Science. Her M.Ed. was awarded from The University of Florida. She taught biology in the Austin, Texas public school system and now teaches science at Westwood Middle School in Gainesville, Florida, where all the exercises in this manual were developed and tested.

TABLE OF CONTENTS

Introduction

Science is fun!!! Students learn best about the enjoyment and excitement of science by becoming involved in the scientific process. We developed this series of hands-on exercises as a way for students to **do** science, rather than just **learn about** science. These hands-on activities were designed to allow all students to be involved in self-discoveries about life and, in particular, plant life. While students learn a lot about plants, the more important lessons are about discovery and the scientific process.

While students learn a lot about plants, the more important lessons are about discovery and the scientific process.

These exercises were developed to supplement normal classroom activities by allowing students to initiate ongoing projects to investigate the biology of anatomy, development, and physiology. In most cases these exercises are open-ended. This manual provides information to allow teachers to introduce a topic and to initiate the exercises. The end point is not fully predictable for any exercise or student group. Because there are no negative results in science, the student's challenge is to understand the results that they do obtain. Instead of the original objective for an exercise, students may well learn about the growth of fungi or the effect of a stress on plant development.

As students begin thinking about each exercise and the results they obtain, they learn that discussions commonly end with "I don't know." For the teacher and the students this is an exciting point. This is when the imagination and inspiration of science is learned. Students are then challenged to search for more information about the topic, to develop alternate hypotheses, to explore the unknown, or to plan further investigations to explore the problem.

Importantly, these exercises were introduced into our middle school curriculum without any extra financial resources. Therefore, each exercise was developed so that it could be implemented with little or no added expenditures. The essential "equipment" required for each exercise is commonly constructed by the students from recycled materials brought from home, such as styrofoam trays, small glass jars, and soda bottles. Some required materials may be donated or purchased at a discount price from a local plant retailer or florist. (In fact, these exercises provide an excellent context in which to invite these people to the classroom to discuss the importance of biological knowledge in their business.)

Nature of the Exercises

We present 14 exercises we have used to supplement the usual classroom activities in an introductory biology class. Each of the exercises is essentially independent and should be selected to fit the curriculum and facilities of each classroom situation. It is anticipated that each teacher will find ways to improve on these exercises to better fit their specific situation.

These exercises should not be viewed as a rigid curriculum in which each exercise must be used and presented in the sequence presented in this manual. While there is some building of vocabulary by using the sequence presented here, each teacher should consider how to best integrate selected exercises into existing curricula. Further, we expect a teacher may initially choose to use only a few of the exercises during a school year to complement classroom activities. In fact, the exercises themselves may lead to other classroom activities that expand the time requirements. Even though we have used each of these exercises several times as ongoing classroom activities, we have never had the time to schedule all 14 exercises in one year.

While there is some building of vocabulary by using the sequence presented here, each teacher should consider how to best integrate selected exercises into existing curricula.

Each exercise can extend over a period of one to three weeks. Initially, a full class period (43 minutes in our school) may be used to introduce the topic, to allow students to enter procedures into their notebooks, and to begin the exercise. Usually during the next one to three weeks, students on a regular basis will need to make observations on the exercise and record the results. In most cases, this requires only 5 to 15 minutes of a class period. We commonly devote a full class period at the end of an exercise to make final observations, discuss results, and allow students to write their conclusions about their results.

Exercises are usually done in groups of two to four students. Of course, students enjoy the chance to socialize and to work together on projects. Important to learning about science, group work provides a check on what is being done and facilitates discussion. Thinking about and challenging ideas in science is very much an interactive activity; working in groups fosters this concept.

In our use of these exercises, each student maintains a notebook as an essential part of the activity. We have asked students to maintain complete records of the procedures, results, and conclusions. Each student approaches observations differently so we encouraged drawings, graphs, special observations, and general comments as legitimate methods to record results. Finally, at the end of an exercise, students include in the notebook their discussion and conclusion from their results. These conclusions are to include ideas of applications of this new information or a proposed "next step" in the study. Keeping journals and good records are important in many activities, and they are essential in science.

Orderliness and completeness are emphasized in maintaining the notebooks in order to allow students to retrieve information. Rather than ask students to memorize vocabulary and concepts that come from the exercises, we have them record this information in the notebooks. Open-notebook quizzes are a way to affirm the importance of good record keeping and to reinforce some of the ideas that may have been learned in an exercise. We specifically chose not to require memorization of exercise material, but rather we encourage using the notebook as the key resource for learning information.

Rather than ask students to memorize vocabulary and concepts that come from the exercises, we have them record this information in the notebooks.

Content of Teacher's Manual

This manual describes 14 hands-on exercises that we have used in a middle school introductory biology course. We have attempted to give full details about the operation of each exercise in our classroom. However, it is fully expected that the exercises may need to be modified by teachers to match their own interests, the needs and capabilities of their students, the classroom format, and existing facilities and equipment. In this manual we attempt to present what has worked for us in sufficient detail so that teachers can initiate the exercises with their students.

There are seven parts in the description of each exercise.

(1) *Education objectives* are listed because the exercises commonly provide learning opportunities beyond the obvious investigation of a biological phenomenon.

(2) *A brief biology background* is given for those teachers who are not fully familiar with plant science. It is hoped that sufficient information is given so that a teacher can initiate discussions on individual exercises.

(3) *A description of the experimental environment and materials required* for each exercise is presented so that a

Rather than using handouts to be put in the notebook, we chose to put material on the blackboard for students to use.

teacher can assess if the resources are available to initiate a particular exercise.

(4) *A detailed presentation of the procedures* is given for the teacher. Sufficient detail is attempted so that the teacher can evaluate the commitment of time and resources required to undertake the exercise.

(5) *Suggested information* is included for students to use in developing their notebooks. At least at the middle school level, we found students needed some guidance on the important records to be kept in their notebooks. Rather than using handouts to be put in the notebook, we chose to put material on the blackboard for students to use.

(6) *Topics to stimulate discussion* and critical thinking among the students are given.

(7) *Ideas for further activities* are listed on each topic for use by the whole class or by students that might want to pursue independent studies.

FINAL COMMENT

Science is fun, but it also requires careful work and an expanding knowledge. Students at the introductory level should be challenged by these exercises but not overwhelmed. Every exercise yields interesting results even if they are not those that were anticipated. The challenge for the students is to solve the mystery of what happened and, if possible, why. It is imagination in answering these questions that is the essence of science. We hope teachers are rewarded, as we have been, in having students develop new interests and skills in their thinking about biology. Our greatest acclaim is to have students tell us at the end of the year that biology is "cool."

FLOWER ANATOMY 1

BACKGROUND

Educational Opportunities for Students

A. Learn the importance of critical observation of "ordinary" things.
B. Learn the need to keep a detailed notebook of your work.
C. Learn laboratory safety in dissecting biological specimens.
D. Learn the parts of a flower.

Biology Background

Many students are introduced to the classical "scientific method" as the way in which science is done. While this is an important idea for students, in practice the key to most science is careful observation of what is before you. In the history of science, many of the important discoveries have occurred simply because someone undertook critical and careful observations of events or things going on around them. Therefore, an important lesson for the students is to develop a sense for making critical observations of something that they have looked at many times — for instance, a flower.

Another significant part of scientific discovery is the need to keep complete records of observations. (In fact, careful record keeping is an important part of almost any profession.) Students are to keep a record of their experimental work in notebooks by entering the procedures, experimental data, drawings, impressions, and any general observations. Students should be encouraged to write just about anything they choose in their notebooks. (A scientist usually learns quickly that you cannot put too much in notebooks.) Notebook neatness and presentability to others are <u>not</u> important criteria for good notebooks. Teachers need to check notebooks for completeness and the inclusion of student impressions and conclusions about each exercise.

Flowers come in virtually all shapes and sizes. For the first day of this exercise, it is assumed the teacher will provide the flowers to the students. It is good to use a complete flower containing readily visible male (*stamen*) and female (*pistil*) organs. Gladiolus flowers are large flowers in which various organs can be easily identified. After identifying the major components of the flower, encourage the students to dissect the flower, looking for detailed parts of each organ. This is an exploratory exercise so students should use the flowers to become familiar with using lens and dissecting microscopes, if available. The students should be able to identify *pollen* and *ovules*.

A second day of the exercise could include observations of any flowers the students can find and bring to class. Unquestionably, a wide range of flower types will be brought to class. Petal coloration, number and fusing of stamen, and split stigma of the pistil are all possible variations. Also, some "flowers" are imperfect and contain only the female or male organs. Again, this is an exploratory exercise so in some cases the students (and teacher) will simply have to give their best guess. Remember, the interpretations of plant anatomy are not static as scientists develop more sophisticated equipment and techniques for observation.

Materials and Experimental Environment

- Complete flowers, at least one for each lab group (stalks of gladiolus flowers, usually available from a florist, are good; there will be variation in development on a stalk)
- Single-edged razor blades
- Hand lenses and/or dissection microscopes (binocular)
- Fresh or dry flowers brought in by students (for second day of study)
- Textbooks or reference materials

Procedures

The introduction to this exercise should include a discussion of laboratory safety. In particular, students should know that **special care is required in using razor blades**. The students need to be aware of laboratory procedures for obtaining the razor blades, their use, and their disposal at the end of the exercise. Students should be shown that the plant material is placed on the dissecting surface and held firmly while gently cutting through the tissue. While we never have had an accident, procedures for dealing with an accident should be presented to the students.

In this activity, students will be drawing and labeling from their own specimens.

1. From a gladiolus stem, cut an open flower for each lab group. As the flowers on a stem will be at different stages of development, encourage students to compare specimens.

2. Students will examine and draw what they see, labeling these key components: petals, pistil and ovules, stamen and anthers. Textbooks and other reference materials should be available and students should be encouraged to compare labels with them as well as with drawings on the board.

3. Using single-edged razor blades, students can cut longitudinally through the ovulary and examine the ovules with hand lenses or dissecting microscopes, drawing what they see. Lab groups should compare specimens, as several stages of ovule development will be present.

4. Unopened flowers should be examined in the same way.

5. Students should also examine and draw anthers and pollen grains, using hand lenses and microscopes, and compare specimens as before.

6. A second day of this exercise can be an examination of fresh or dried flowers brought in by students. Students should observe as many different flowers as resources allow.

NOTEBOOKS

Blackboard Information

Flower Anatomy	DATE: ____________________

OBJECTIVE: To discover the different parts of a flower.

PROCEDURES:

1. Obtain flower from teacher.
2. Draw flower and label major parts.
3. Carefully dissect parts with razor blade.
4. Examine parts with lens and microscope.
5. Draw small parts and label.

Flowers II	DATE: ____________________

OBJECTIVE: To explore the anatomy of more flowers.

PROCEDURES:

1. Draw flower and identify major parts.
2. Carefully dissect male and female tissue.
3. Draw and identify pollen and ovules.

Results

Students will make drawings of the entire flower and its components. We suggest that you put some helpful sketches such as the following on the board. However, emphasize that sketches are for reference and that students should draw what they have seen for themselves.

FIGURE NO. 1 COMPLETE FLOWER

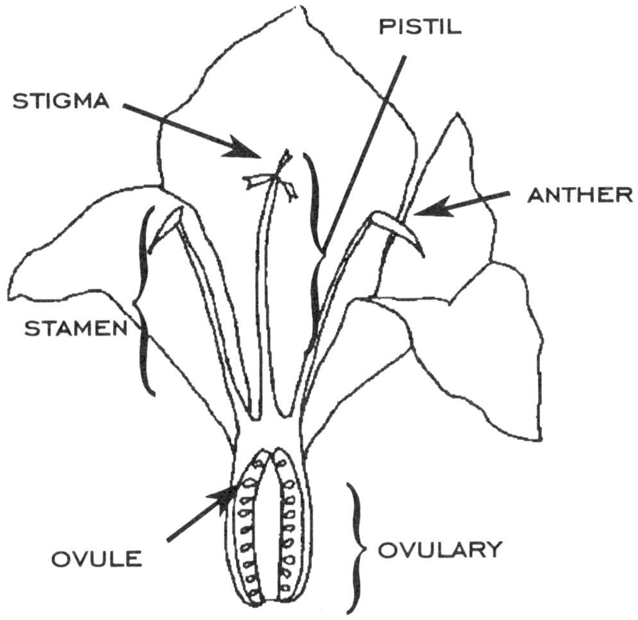

Discussion Ideas

1. What have we learned?

2. What is the importance of the flower to a plant?

3. What is the purpose of the following: stigma, stamen, ovulary, anther, pollen grain, ovule?

4. What differences between flowers have you noticed?

5. Can you think of reasons for flowers not all being alike?

Further Activities

1. Conduct a wide-ranging comparison of the anatomy of many flowers. This study could include the *inflorescence* (i.e., reproductive organs) of plants without "flowers." Trees and grasses offer challenging anatomical studies.

2. Observe a series of anatomical changes of flower development from the very earliest stages to full flower maturity.

3. Collect pollen from different flowers and compare anatomy. Pollen anatomy has been important in many archaeological and forensic studies.

SEED ANATOMY 2

BACKGROUND

Educational Opportunities for Students

A. Learn to use the notebook as an important tool for recording observations and insights.

B. Learn about the embryo stage of life, and about its importance as a quiescent stage in the life cycle of plants. Eating all your favorite seeds means you are eating "baby" plants — look for them!

Biology Background

Seeds are crucial in the life cycle of plants because they are the means of regenerating the species. Seeds are structures to carry the species through environments that would threaten the species with extinction in the vegetative stage. Freezing temperatures of winter and lack of water in drought periods would result in the extinction of many plant species if not for the great hardiness of seeds.

To perpetuate the species, seeds must contain the *embryo* of the next generation of plant and the "energy" reserves on which the developing embryo can grow. In many seeds the embryo is reasonably large and readily identifiable. Peanut and sweet corn seed are familiar to students and have embryos sufficiently large for identification of vegetative parts. The students should be able to locate and identify the *radicle* or developing root, and the *plumule* or developing shoot for each seed type.

The storage of energy reserves in the seed is crucial, and much of the volume of the seed is made of storage materials. (In Exercise 10, the students will investigate a bit more about the nature of these storage materials.) The *cotyledon* is the structure that is involved in supplying the nutrients to the developing seedling. The number of cotyledons is used to classify plants. Monocots have one cotyledon;

dicots have two cotyledons. Grasses and palms represent two major groups of monocots.

In peanuts, the two large cotyledons are easily identified as the "two halves" of the peanut seed. In sweet corn, the cotyledon is more difficult to identify because it is located adjacent to the embryo and has much the same appearance as the embryo. (Exact identification of the cotyledon in sweet corn is not a key point for the students.) Sweet corn seeds have a large *endosperm* that contains the storage materials for the developing seedling. In the case of sweet corn, the cotyledon helps to process the materials in the endosperm for use by the developing embryo.

The *seed coat* should not be overlooked as an important structure of the seed. Seed coats provide an important barrier against diseases (i.e., microbes), insects, and in some cases, digestion in animal intestines. The peanut seed coat is the red film that comes with roasted peanuts. The seed coat of interest to students may be the one on popcorn. The popcorn seed coat is essentially air tight and physically strong. Upon heating, the water inside the seed is vaporized and causes gas pressure in the seed to increase. Increasing gas pressure causes the seed coat to eventually rupture and the starch of the endosperm is exploded outward. (If the students have studied gas pressure, they could be given a few hints and then challenged to explain the popping of corn.) The "husks" that get stuck in your teeth are the ruptured seed coats.

Materials and Experimental Environment

- One roasted peanut for each student (in shell, if possible)
- One whole seed of sweet corn for each student (If sweet corn seeds are unavailable, field corn seeds soaked for 24 hours will work.)
- Various seeds brought in by students
- Metric rulers
- Single-edged razor blades
- Hand lenses and/or dissection microscopes (binocular)
- Popcorn
- Popcorn popper (optional)
- Textbooks or reference materials

Procedures

This exercise should be preceded with a reminder about laboratory safety. If the students have completed the flower exercise, then they need only a reminder to dissect the seeds by placing them on the dissecting surface and to hold the seeds firmly when cutting. The corn seed will require more pressure to cut than the flower and, hence, more care in cutting. In addition, seeds may have been pretreated with a fungicide, so students need to get in the habit of washing their hands during and after laboratory exercises.

In this activity, students will be drawing and labeling from their own specimens.

1. If roasted peanuts in shells are available, students can investigate and draw the shells first, including measurements of its dimensions. Call attention to the "bumps" on the shells. Have students break the shell and gently tear the shell apart. They will discover that the "bumps" correspond to stringy tissue that will separate out with some practice. These strings are vascular tissues (*xylem* and *phloem cells*) that carry nutrients for the shell (that is, *pod wall*).

2. Students will examine and draw an unopened peanut seed, including measurements of its dimensions. Then, they will gently open the seed along the "seam," drawing and labeling what they find inside. The two cotyledons make up most of the seed. The embryo, comprised of the root and shoot, should be readily apparent and labeled in the students' drawings.

3. On the second day, students will examine, draw, and dissect a corn seed. (If students are starting with seeds purchased from a seed store, the seeds may be coated with a fungicide [usually a pink powder] which is poisonous to humans. Warn students to avoid putting their fingers in their mouths. They should wash their hands after handling treated seeds.) The black tissue at the tip of the corn when opened is the spot where the seed was attached to the mother plant. Students should place the corn seed on the dissecting surface with the white area up. The ridge in the middle is the embryo and the remaining white area is the cotyledon. The bulk of the seed is an endosperm (which doesn't occur in dicot seeds). A vertical cut through the length of the embryo should split the seed for more detailed observations.

4. A third day of this exercise can be used to examine seeds brought in by students. Remind the students that seeds are

everywhere. Soaking most seeds for 24 hours before the class will usually make dissection much easier. Students will examine and dissect seeds and record as much information as they wish in their lab book.

5. For homework, have the students find out what happens when popcorn pops. In your windup discussion of seed anatomy, discuss their answers and then pop some corn for the class.

NOTEBOOKS

Blackboard Information

Peanut Seed Anatomy DATE: ______________________

OBJECTIVE: To discover the parts of a peanut seed.

PROCEDURES:

1. Examine and draw a peanut pod.
2. Gently tear the pod and observe the tissue.
3. Gently open the seed along the "seam."
4. Examine the structures exposed and draw and label them.

Corn Seed Anatomy DATE: ______________________

OBJECTIVE: To discover the parts of a corn seed.

PROCEDURES:

1. Examine and draw a corn seed.
2. Place the corn seed white-side-up and cut the seed lengthwise through the middle.
3. Draw what you see and label what you can.

Results

Students will make drawings in their notebooks. We suggest that you put some sketches on the board as a reference (see next page). However, students should draw what they have seen for themselves.

FIGURE NO. 2 HALF OF PEANUT SEED

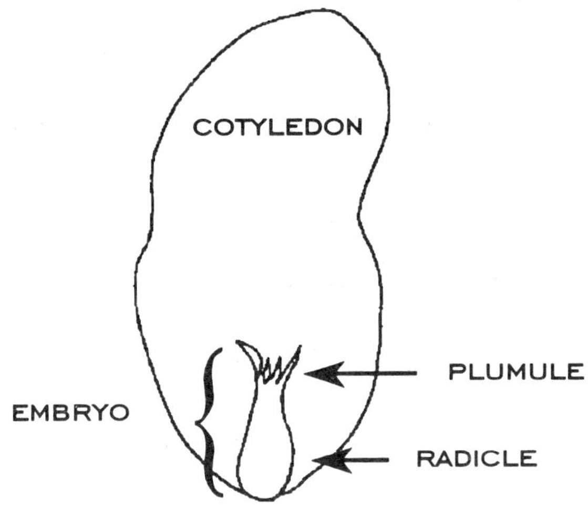

Discussion Ideas

1. What have we learned?

2. What is the importance of the seed to a species?

3. Why is the embryo in a seed instead of somewhere else on the plant?

4. Give some examples of threats to seeds.

5. What is the importance of the cotyledon to the developing seed?

6. What does the seed coat do for the seed?

7. What happens when popcorn pops? How is popcorn different from regular corn?

8. When we eat popcorn, what is the "white stuff"? What part of the seed gets caught in our teeth?

FIGURE NO. 3 CROSS-SECTION OF CORN SEED

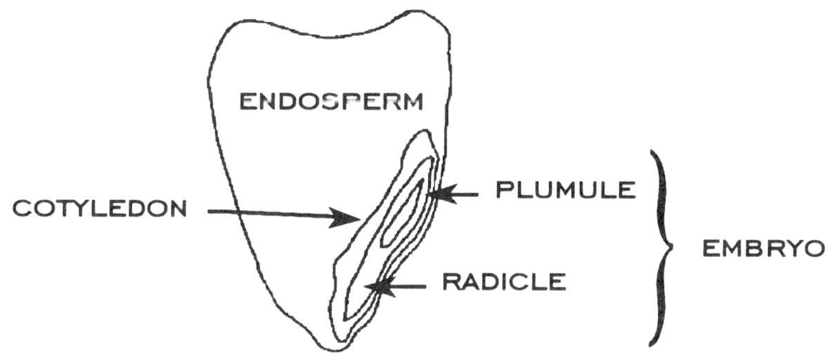

Further Activities

1. Make a collection of different seeds and compare their anatomy.

2. Try to identify the components of more exotic seeds such as coconut or strawberry.

3. Obtain a soil sample and carefully search for seeds. A soil sample from an old crop field or garden on which weeds had grown abundantly would contain many seeds. Many seeds are small, so take considerable care.

SEED GERMINATION 3

BACKGROUND

Educational Opportunities for Students

A. Learn about organizing an experiment and maintaining a continuing record of the results.
B. Learn about the germination process in seeds.

Biology Background

Seeds move from their quiescent stage with the *imbibition* of water. Imbibition of water occurs in the presence of adequate liquid water, and is the process where water penetrates the seed coat and the cells of the seed begin to swell. Seed swelling is dramatic and will be the first event documented by the students.

The imbibition of water triggers many biochemical events in the seed. Storage materials in the cotyledon and/or endosperm are mobilized and made available to support the growth of the embryo. Usually the radicle (i.e., the embryonic root) first begins to grow and breaks through the seed coat. The growth of the radicle is important in the absorption of water to sustain continued growth of the seedling. (In Exercise 8, downward radicle growth is examined in more detail.) The plumule or shoot usually emerges from the seed after the radicle.

The exact positioning of growth along the radicle is very important in determining the positioning of the cotyledons on the developing seedling. If the growth is only at the tip of the radicle, then the radicle pushes deeper into the soil and the old seed remains in its original position. This type of germination is called *hypogeal* and is the type of germination exhibited by corn and peas. If the radicle also elongates near where it is attached to the cotyledon, then the cotyledons are pushed above the soil. This type of germination is called *epigeal* and is the type of germination for many beans.
In epigeal germination, the cotyledons usually unfold once they

have emerged from the soil. Chlorophyll is synthesized in the unfolded cotyledons (see Exercise 11). Therefore, the green cotyledons can begin photosynthetic accumulation of "sugars" by the seedling at an early stage so that rapid growth is promoted. The disadvantage of epigeal germination is that it exposes the storage materials in the seeds above the ground. Therefore, severe damage by cold temperatures or predators is more likely for species with epigeal germination than those with hypogeal germination.

The germinating seed is vulnerable to microbe infection, particularly fungi. Microbes grow abundantly when water is plentiful and the compounds in the seeds provide the resources for growth. Seeds are commonly treated with chemicals (fungicides) to inhibit microbe growth in the vicinity of the germinating seed. By germinating seeds in a very moist environment such as used in this experiment, it is likely that even with chemically treated seeds some seeds will become infected with fungi. Student observations on fungi are an opportunity to learn of difficulties in doing experiments and to observe another life form.

Materials and Experimental Environment

This experiment should be started on a Wednesday or Thursday so observations can be made the day following the start of the exercise and during the subsequent week.

- Seed germination chambers, one for each lab group. (The germination chambers are constructed by the students from styrofoam meat or produce trays. A good size is approximately 15 cm by 25 cm. Hinge two of the trays together with masking tape as in the drawing on the next page.)
- Seeds of corn, pea, and bean
- Paper towels
- Styrofoam trays
- Masking tape
- Metric rulers
- Shelf area to store germination chambers

Procedures

1. Each lab group constructs a germination chamber using styrofoam trays and tape. The trays should be labeled with

FIGURE NO. 4 SEED GERMINATION CHAMBER

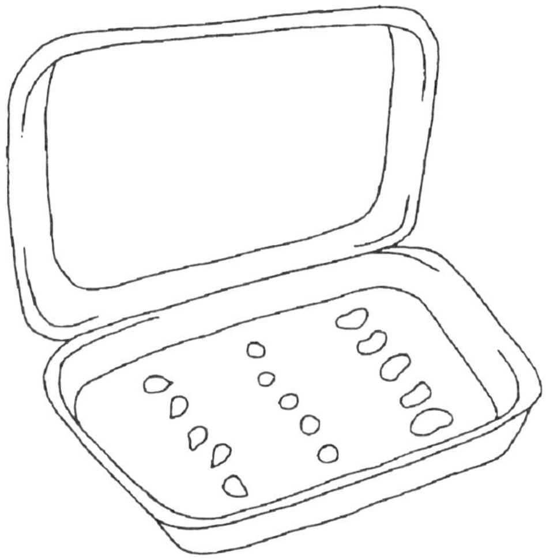

each group's identification. Chambers are prepared by putting two water-soaked paper towels in the bottom tray. Adding a small amount of water (depth < 1 mm) may be helpful.

2. Five seeds are selected from each species. Lengths and widths should be measured for each seed and an average calculated for each type.

3. The seeds are placed in the chamber and covered by a third, damp paper towel. Students should take care to keep the paper towels from sticking out of the chamber and contributing to drying. Chambers are taped shut and left overnight. We found it useful to store the chambers for each class together in a preparation room to avoid mix ups among classes.

4. Beginning the next day, students will check the seeds daily, measuring seed size. As germination progresses, the length of the developing radicle and shoot are also measured.

   Students should record any observations under comments for later discussion. Many things can happen. Seeds fail to grow or fungus develops — this is the beginning of the realization that experiments don't always go the way they were planned.

5. Graphing is important in visualizing experimental results. It is assumed that the students have already been introduced to graphing. After finishing the experiment, students should graph the results. Many of our students were confused when asked simply to graph the results. They should be challenged to review the type of graphs and select the one that would work best to visualize their results. A line graph with time as the independent variable (x axis) and length of various tissues as the dependent variable (y axis) will allow them to readily visualize the changes in tissue dimensions.

NOTEBOOKS

Blackboard Information

Seed Germination DATE: ____________________

OBJECTIVE: To observe different types of seed germination.

PROCEDURES:

1. Construct a germination chamber from two trays and masking tape. Put masking tape label on tray.
2. Select five seeds each of corn, pea, and bean.
3. Measure and record the mean (average) length of each type of seed.
4. Place two paper towels in the bottom of the chamber and add water to less than 1 mm depth.
5. Arrange the seeds on the wet paper towels, then cover with a damp paper towel. Tuck all towel edges inside the tray and close the chamber with tape. Label the tray.
6. On a daily basis, observe and record changes, and if needed, add water.

Results

Students should prepare a table like the one shown below in their notebooks for each type of seed.

Date	Lengths (mm)			Comments
	Seed (cotyledon)	Exposed Radicle	Exposed Shoot	

Discussion Ideas

1. What have we learned?

2. What is imbibition? What physical process accomplishes this? (osmosis)

3. What does imbibition do for the plant? How do you know?

4. What part of the embryo showed up first? Does this make sense? Why?

5. What is an advantage of epigeal germination? A disadvantage?

6. What plants show epigeal germination? Hypogeal germination?

Further Activities

1. Germinate different kinds of seeds.

2. Carefully mark different segments of the radicle and/or shoot with a marking pen. Observe if the distance between adjacent marks increases to indicate where the tissue is growing.

GERMINATION & ENVIRONMENT 4

BACKGROUND

Educational Opportunities for Students

A. Introduce students to the "hypothesis" approach to scientific investigation. Stimulate student thinking about experimental variables, the need for a "control," and types of data to be collected.

B. Introduce the concept of "do-able" experiments to the students. Brainstorming should produce some interesting ideas from the students for experiments. Then the ideas must be refined to the do-able level. That is, experiments must be adjusted in view of the resources and knowledge available to do the experiment.

Biology Background

Many variables in the environment surrounding germinating seeds have important effects on germination and seedling development. In fact, a number of conditions have to exist within a fairly narrow range to achieve successful germination. Many of the treatments students develop will probably have very negative effects on germination compared to controls. These negative responses can be quite dramatic and exciting to the students.

The brainstorming by our students resulted in many imaginative ideas. Many ideas seemed "silly" and were outside the do-able limitation, but these too were valuable starting points. Proposed ideas for an experiment on drug effects on seeds became a test of aspirin in the germination water; an experiment on alcohol effects became a test of cola in the germination water; and a test of winter survival became an experiment on the effect of seed storage in a freezer before the germination test. The concept of keeping it do-able by studying only one or two variables within the allotted time-frame of a couple of weeks is an important challenge to students.

The "do-able" issue confronts scientists in virtually every experiment they design. The students also need to consider the nature of the control, and what actually is the control for the specific treatment being investigated.

The outcomes of the various experiments are truly unknowns. Because of the nature of these student experiments, many are likely to have never been performed. The students must use their own minds in obtaining the results and in developing their own conclusions. Probably, the teacher's role is to probe the students with questions about alternative conclusions, the implications of their experiment in view of their original hypothesis, and importantly, ideas for improving or repeating the experiment. Remember: Some of the most significant knowledge in science is gained when things go "wrong"!

Materials and Experimental Environment

Experiments should probably be started on a Wednesday or Thursday so that all of the following week can be used for observations.

- Seeds
- Germination chambers
- Metric rulers
- Other materials, depending on variables chosen (thermometer, chemicals, etc.)
- Shelf area to store germination chambers

Procedures

In this activity, students will design and organize their own experiments.

1. Begin the exercise with a class discussion. Point out that, up until now, their findings have been observations. Now, based on their previous observations, they can do hypothesis testing. Have them brainstorm environmental factors that might affect germination (e.g., soak/no soak, light/no light, hot/cold, etc.). Discuss general considerations such as accurate measurements, intervention of weekends, controls, and duration of experiments. Remind them to ask themselves, "Is it do-able?"

2. After discussion, the lab groups meet to choose and design their experiments. No variable need be rejected unless it is

dangerous, expensive, or otherwise impossible. Groups should consider each idea in terms of control vs. treatment, how the treatment will be done, and what will be measured as well as when and how. It is likely seeds from only a single species need be tested per lab group. The groups should keep in mind that they will be responsible for providing materials not available in the classroom (such as oil, tabasco sauce, or hair spray). Once the decision is made, each group should formulate a hypothesis, agree on materials and methods, and record these in each individual's lab notebook.

3. After allowing a few days for finalizing experiments and organizing materials, lab groups will set up their experiments and make initial observations. Germination progress will be measured over time, according to the design of the experiments.

4. When the experiments are finished, groups will organize data and draw conclusions. This is a good opportunity to practice oral reporting of the studies as well as recording written conclusions discussed in their lab books.

NOTEBOOKS

Blackboard Information

(Group's choice of variable) and Germination
DATE: ___________

OBJECTIVE:
To test the hypothesis that germinating seeds of (species) are affected by treatment(s) of ______________________________.

PROCEDURES:
(Your procedure and materials here.)

Results

The students will likely need to develop a table to record data. They probably should include at least one graph giving final results and reflecting your hypothesis.

Discussion Ideas

1. What have we learned?
2. Why do we have a control?
3. What is a variable?
4. How many variables can you/should you test in one experiment?
5. What are we considering when we ask, "Is it do-able?"
6. Is it possible to write too much in your lab notebook?
7. What is the difference between a final presentation and the lab notebook?
8. React to the statement, "No one has ever done the same experiment my group has just finished."
9. Is an experiment "gone wrong" a waste of time and material?

Further Activities

1. Did the experiment indicate other factors that need to be included? Does an alternate control need to be tested?

2. Many ideas could likely be expanded into a "dose response" study. That is, try several different levels of the treatment.

3. Are there combined treatments that might interact to affect germination?

PLANT GROWTH & ANATOMY 5

BACKGROUND

Educational Opportunities for Students

A. Confront the problem of growing green plants.
B. Learn general anatomy of two types of plants.

Biology Background

Many exercises do not work well unless the experimentalist devotes considerable care and effort to them. In this experiment, students will construct their own pots and grow plants from seeds. Many things are likely to go wrong: improper water drainage from pot, undesirable soil, seed sowing depth too deep or shallow, inadequate watering, and insufficient light. Recognition of problems that may develop in the growth of these plants is an important learning exercise. Good note-taking will help the students interpret the performance of their experimental techniques and help them improve on these techniques in subsequent experiments. (If nearly all students fail to grow plants, then this experiment probably should be repeated.)

It is proposed that two types of plants be grown. A bean typifies the growth of a dicot plant while corn typifies a grass plant. Successful growth of a single plant of each plant type in a class is sufficient for students to make drawings of the plant. At two to three weeks old, the plants of the two types will look considerably different. The bean will have a long stem with new leaves clearly developing at the stem apex. The corn does not elongate a stem until a number of weeks into growth. The apical meristem for the corn remains below the soil and new leaves push up in the whorl of the older leaves.

The leaf type is quite different between the dicot and grass. The leaf of the bean is composed of three leaflets, making it a compound leaf. The vascular bundles form a complete network

FIGURE NO. 5 CORN SEEDLING

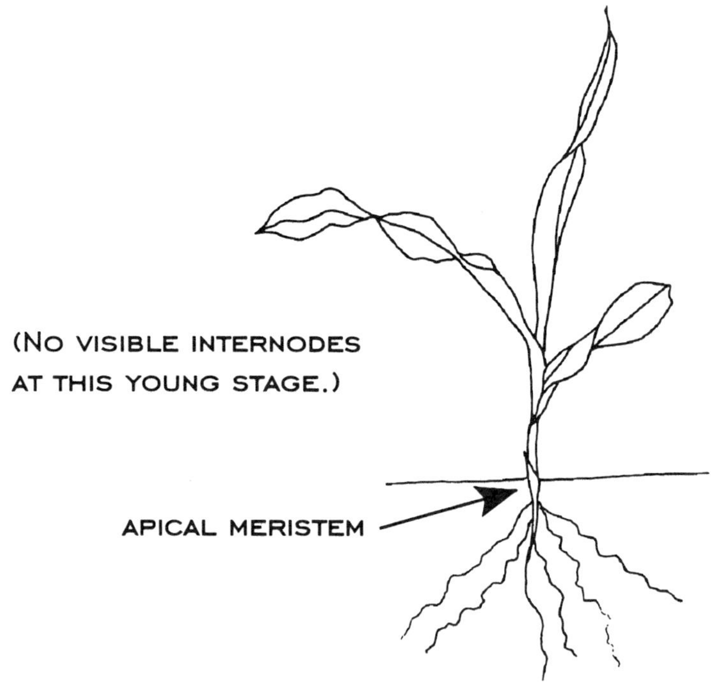

throughout the leaflets. The leaf of grasses is "simple" with parallel vascular bundles. Because the meristem for grass leaves is at their base, if you remove the end of the grass leaf it will continue to grow. (This difference in the placement of the meristem of leaves allows a further experiment: following what happens when tips of leaves of the two leaf types are removed.) Students who mow lawns will be enthused to learn that the placement of the stem apex below the soil and of the leaf meristem at the leaf base allows lawn grass to be frequently mowed without damage.

Root patterns are also quite different between the two plant types. The bean will develop a dominant, deep primary root. In some dicots, this trait is used in arid regions to seek water reserves in the soil at very great depths. Grasses usually develop a more "fibrous" root system where the soil is permeated with a network of roots of similar size. This allows the grass to fully exploit the materials stored in the soil.

FIGURE NO. 6 BEAN SEEDLING

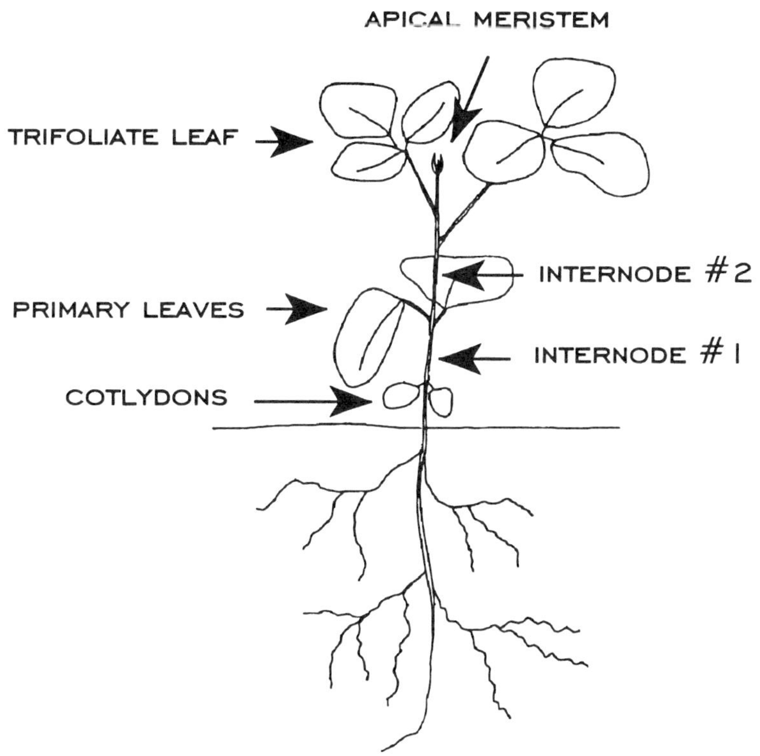

Materials and Experimental Environment

- Corn and bean seeds
- Styrofoam cups
- Potting soil
- Metric rulers
- Hand lenses and microscopes (binocular)
- Sunny location to grow plants

Procedures

In this activity students will construct their own pots, grow plants from seeds, keep records on plant development, and learn the general anatomy of dicots and monocots.

1. Each lab group will prepare two pots for planting by making three 3-4 mm drainage holes in the side of a styrofoam cup—very close to the bottom of the cup.

2. The cup should be filled nearly to the top with gently packed potting soil.

3. Each cup should be sown with three or four seeds of a single species. The seeds should be sown at a depth of about 2 to 3 cm. The students should then set their pots on trays (cafeteria trays work well) and put them in a sunny spot. We used the top of book shelves built under south-facing windows in the classroom for growing the plants. Watering should not take place until the pots are on trays. This will keep potting soil out of sinks and water off the floor.

4. Students will make regular observations over a two- to three-week period. Data to be recorded include stem length, leaf size and shape, and position of the cotyledons. Students will need assistance in observing corn because the meristem stays below ground in the seedling stage, and there will be no stem length to measure. Also, discussion about what is a node and an internode of the bean stem will be helpful. Encourage students to leave ample space in the chart for comments, reminding them that too much information is better than too little.

5. When the final measurements are made, students can remove the plants from their cups and lay them out on a paper towel for comparison and sketching. Remind them to consider the type of germination (epigeal vs. hypogeal), differences in roots, and meristem location. Groups who have had poor success with growing plants will be able to look at the plants of other groups for final comparisons.

6. Developing graphs in their notebooks is, again, a very effective way for students to summarize and visualize their observations. Line graphs of length plotted against time is likely to be the most effective presentation of this data.

NOTEBOOKS

Blackboard Information

Plant Growth and Anatomy DATE: ____________________

OBJECTIVE: To grow and examine plants of corn and bean.

PROCEDURES:

1. Make two pots.
   a. Label two styrofoam cups.
   b. Put three 3-4 mm holes in the sides at the bottom.
   c. Put in soil and gently pack.
2. Sow three seeds of each type in different cups at 3 mm depth.
3. Put in growth location and water.
4. Observe length of stem, leaf number, and leaf for approximately two weeks.
5. Water as needed.
6. After two weeks, remove plants from pots. After gently removing soil, lay one of each type on a paper towel.
   a. Make final measurements.
   b. Draw each species and note differences. (Consider epigeal or hypogeal germination, leaf number and shape, differences in roots, and meristem location.)

Results

Students should prepare in their notebooks one table for each species as shown on the next page. Drawing the plant will help students reference data.

Date	Length (mm)						Comments
	Internode #1	Leaf #1	Internode #2	Leaf #2	Internode #3	Leaf #3	

Discussion Ideas

1. What have we learned about growing plants from seed?

2. How do the three plants differ in: where the leaves grow from, leaf number and shape, pattern of roots, epigeal vs. hypogeal germination?

3. What is the role of the hairs on leaves, roots, stems?

4. If you were planning a new experiment, which of the two species would you like to use? Why?

Further Activities

1. Sow other species and observe their development pattern.

2. Grow plants into latter stages of development to observe the development of additional leaves and roots. If growth facilities are available, plants could be grown to reproductive stages.

3. Observe range of plant structures that exist in natural settings or in domesticated plants. A visit to a local plant nursery will provide an opportunity to observe in detail a wide diversity in plant structure.

HORMONES & ASEXUAL REPRODUCTION

6

BACKGROUND

Educational Opportunities for Students

A. Increase observational skills and confidence in working with plants.
B. Learn about regulatory role of hormones in influencing development.
C. Learn basic concepts of asexual reproduction and cloning.

Biology Background

Every living cell in an organism contains the complete DNA code to reproduce the entire organism. Cells mature, the DNA within the cells replicates, and the cells split to form two new daughter cells (*mitosis*)—each with a complete set of DNA for regenerating an identical organism. Although the daughter cells have the DNA code to create the entire organism, these daughter cells develop as prescribed by their position in the organism to contribute to the existing individual. One of the great advances in evolution was the regulation of the DNA code so that, in any cell or at any given stage of development, only a small relevant part of the DNA message is active.

One of the achievements of modern biotechnology has been to extract individual cells from a parent organism and "convince" these cells to begin development in isolation of an entirely new organism. That is, these extracted cells are tricked with hormones and other signals into behaving as if they are zygotic cells of a new individual organism. When successful, these cells continue to develop to produce an organism that is genetically identical to the parent organism. This procedure of replicating the parent organism is called *cloning*.

The procedure of cloning from individual plant cells is still limited to a few species and is complex and difficult. However, new plants can be cloned from plant parts in a number of species by asexual reproduction. In asexual reproduction, a cutting of a plant is taken and the development pattern of individual cells in the cutting is altered to form a new plant. In this experiment, cuttings of *Coleus* are either untreated or treated with hormone and are placed in water. The development of some of the cells at the base of the leaf stem (petiole) is altered to form roots because of their new relative position. Therefore, after one or two weeks students will visually notice the development of roots from the base of the stem cutting. *Coleus* cuttings readily form roots so the effect of the hormone treatment for this particular species may be minimal. Transplantation of the cutting into soil will allow continued development into a new individual that is genetically identical to the parent plant, that is, a clone.

Asexual reproduction is an important means of reproduction for many plants. It is a method to insure rapid regeneration of plants that are particularly well suited to a specific environment. The danger for the species is that an identical genetic code for all the individuals means that they are all vulnerable to the same external stresses. However, in many horticultural plants, where many of these stresses are controlled or eliminated, the desirability of having identical plants favors the use of asexual reproduction. Some students are likely to have had experiences with cuttings of ornamental plants at home.

Materials and Experimental Environment

- "Mother" *Coleus* plants (white and green variegated *Coleus* plants will be used in Chapter 12), sufficient material for cuttings in each class
- Jars or styrofoam cups, one for each lab group. The styrofoam cups can be transformed to pots once the cuttings are ready to be put in soil.
- Commercial rooting hormone (one container)
- Toothpicks
- Location without direct sun for developing cuttings

Procedures

1. Discuss asexual (vegetative) reproduction. (A good extra credit assignment is for the student to vegetatively reproduce a plant other than *Coleus*.)

2. The teacher should cut for each lab group a 3- to 5-cm length piece of stem from the mother plant to which one leaf is attached at one end of the stem.

3. The students should record the initial length of the stem cutting and the length and width of the leaf. These data may be useful in comparing rooting results between groups.

4. Each laboratory group will be assigned either to leave the cutting untreated or to apply hormone. The end of the stem opposite the end with leaf can be simply dipped into a commercial rooting powder.

5. The cuttings are to be put in a cup of water, propping them in place with toothpicks so that the leaf is out of the water. The cup with the cutting should not be placed in direct sun.

6. Students should observe the cuttings daily and record the appearance of roots.

7. When roots develop to 5 cm or more, students can transplant the new young plants into styrofoam cups. Set the pots on trays (cafeteria trays are good for this) and after a few days put them in a sunny spot.

NOTEBOOKS

Blackboard Information

Vegetative Reproduction DATE: ____________________

OBJECTIVE: To observe asexual reproduction in *Coleus*.

PROCEDURES:

1. Label a styrofoam cup with your name and class period.
2. Put water in the cup to within 3 mm of the top.
3. Obtain a cutting from the "mother" *Coleus* plant. Measure the dimensions of the leaf and stem.
4. If assigned to hormone treatment, dip end of stem without leaf into the hormone powder.
5. Prop the stem piece securely in the cup of water with a toothpick.
6. Place the cup on a tray in the growth location.
7. Observe changes and record measurements daily until ready to transplant.

Results

Students should record observations on the appearance of roots, including estimates of root lengths.

Discussion Ideas

1. What have we learned?
2. What part did DNA play in the formation of a new *Coleus* plant from a small cutting?

3. What was the role of the hormone in influencing the plant development?

4. What could be a drawback to cloned individuals of any species?

5. Why might cloned horticultural plants be more likely to be successful?

6. Have you seen asexual reproduction in plants before? Where?

Further Activities

1. Attempt asexual reproduction with cuttings from other plant species. Treatment with and without hormone can be attempted.

2. A "dose response" experiment with the hormone for various species. Add hormone directly to the water in the cup in differing concentrations and observe root development. Remember, hormones are needed only in very small doses.

TEMPERATURE 7

BACKGROUND

Educational Opportunities for Students

A. Learn the importance of temperature in biology.
B. Expose students to the concept of developing derived data from primary observations.
C. Introduce students to *rate* as an important approach to characterizing performance.

Biology Background

Many chemical reactions and nearly all biochemical reactions are very sensitive to temperature. Within the normal temperature range for life (0 to 40°C), the rate of most biochemical reactions doubles for every 10°C increase in temperature. Consequently, even fairly small temperature changes can have large influences on the growth and activity of organisms. Because most living organisms have little or no capability to regulate their temperature, environmental temperature has a very large effect on biological activity in any particular environment. (The regulation of body temperature at a high, constant level by many animals gives them an important advantage, even though temperature maintenance may require a substantial consumption of energy.)

This experiment allows students to grow plants in two temperature environments that should differ by at least 3°C. Because temperature affects the rate of growth, data should be recorded that allow derivation of rates. Observations on the stem elongation and/or leaf appearance will provide data for the students to plot against time. Graphs of the data against time should allow the students to calculate linear slopes from the graphs to obtain a rate (e.g., mm stem/day). The change of a primary observation with respect to time or another observation is a fundamental concept in scientific data analysis.

Follow up discussions on temperature effects might include a comparison of the forms of life in environments differing in temperature. That is, the importance of temperature can be emphasized in discussing life forms in the arctic vs. tropics or in low vs. high altitudes in mountains. Also, the students could be challenged to think about the consequences of global warming. In many areas, increased temperature will stimulate greater biological activity. Of course, those species that readily respond to the increased temperature will be at a competitive advantage compared to those that have less response. Although not intentionally studied in this experiment, extremely warm temperatures above a species temperature range of adaptation can be lethal. It is anticipated that the geographical range of adaptation for many species may be altered, resulting in potential disruptions of current ecosystems.

Materials and Experimental Environment

- Two (recycled) styrofoam cups for each lab group
- Potting soil
- Bean seeds
- Metric ruler
- Celsius thermometers
- Two locations for growth in cool and warm (heat source) temperature

Procedures

1. Each lab group fills two cups with soil and sows three bean seeds in each pot. This increases the chances of germination in each pot. Plants will be thinned to one plant per pot later.

2. Lab groups should label their pots and place one at room temperature and the other in a heated or cooled environment. Doing this experiment in the winter is likely to mean that the room temperature will be fairly cool. We generated a heated environment by positioning the plants between two thermostatically controlled space heaters. Thermometers should be placed in both environments where they can be easily read.

3. Students will monitor the pots, keeping the soil moist. Remind them to water gently.

4. When plants are about 5-cm tall, the pots should be thinned to one plant per pot. After thinning, observations and mea-

surements every two or three days should be made. Height is measured from the soil up and leaf numbers are noted. Temperatures from both environments are recorded. This experiment should take about two weeks.

5. Graphs in the notebook should be encouraged as a way to determine the growth rate.

NOTEBOOKS

Blackboard Information

Temperature DATE: ____________________

OBJECTIVE: To test the hypothesis that plant development is at a greater rate with warm temperature.

PROCEDURES:

1. Prepare two cups as plant pots. Label the pots.
2. Fill cups with soil.
3. Sow three bean seeds in each pot and water the soil carefully.
4. Place one pot at room temperature and the other in the heated (or cooled) environment. Record the temperature of each environment.
5. Water carefully when necessary to keep the soil moist.
6. After plants are about 5-cm tall, thin each pot to one plant.
7. Observe and measure plant development every two or three days. Record temperatures.

Results

Each student should prepare the following table:

Date	Room Temperature			Warm Temperature			Comments
	Temp.	Plant Height	Leaf No.	Temp.	Plant Height	Leaf No.	

Discussion Ideas

1. What did we learn (what experimental problems did we encounter)? This may include such ideas as proximity of pot to heaters, difficulty in controlling variables, etc.

2. Is there a normal temperature range for life? How do you know?

3. What are the advantages and disadvantages of an organism's ability to regulate body temperature?

4. How can we measure growth? Give some examples other than those we used.

5. How can we relate what we have learned to the possibility of global warming?

6. Could an alien relate the different types of life forms on our planet to temperature zones?

Further Activities

1. If facilities allow, test a greater range of temperatures, or test different species.

2. Examine the "memory" of the temperature treatments. Switch plants between the temperature treatments to determine how quickly their growth adapts to the new temperature.

3. Determine the extreme temperatures at which plants are damaged. Placing plants in a heated cabinet or an old refrigerator for a few hours would allow imposition of treatments. Duration of treatment is another important variable.

GRAVITROPISM 8

BACKGROUND

Educational Opportunities for Students

A. Introduce students to active regulation of tissue development in living organisms.
B. Investigate the importance of gravity in plant development.
C. Introduce use of a protractor.

Biology Background

Gravitropism (sometimes called geotropism in older texts) is a key regulatory process in plants to insure that roots grow down and shoots grow up in developing seedlings. As the student should find, the orientation of seeds in the soil is not important because as the radicle and shoot emerge from the seed their growth is regulated through the action of hormones to cause each organ to grow in the proper direction.

The basic mechanism that results in gravitropism is still not resolved. Sufficiently large particles in cells apparently migrate due to gravity in key cells in the growing tips of roots and stems. The distribution of these particles triggers a number of physiological events mediated by hormones that influence cell expansion. The overall effect on the "gravity side" of the tissue is inhibition of cell elongation in the roots and a stimulation of cell elongation in the shoots. The nonsymmetrical distribution of cell elongation in each of these tissues causes the growing root to bend towards gravity and the stem to bend away from gravity.

An interesting question to pose to students is the problem of growing plants on a space station. This is an issue of considerable current research. Obviously, the directional development of roots and shoots is confused in the absence of gravity. Seeds will apparently need to be germinated under some type of gravity to establish the initial growth of the seedling and then transplanted to growth conditions where other plant controls will regulate the growth pattern.

Materials and Experimental Environment

- Corn seed
- Rectangular containers 7-cm or more deep (Plastic containers in which supermarkets sell fruit work well.)
- Potting soil
- Metric rulers
- Protractors
- Shelf area to store germination containers

Procedures

1. Discuss the experiment, bringing in the question of growing plants in a space station. Assign each lab group one of the three seed orientations (see below). All seeds of each lab group will be sown in the single, assigned orientation.

2. The three orientations of sowing are with respect to the embryo end, or pointed end, of the corn seed.

3. Each lab group selects 18 corn seeds and plants them in three rows, six seeds to a row, according to the assigned orientation. Proper marking of each row (marker, grease pencil) is extremely important, as students will be digging each seed row up on different days. It is easy to forget the direction the rows extend. Take the time to agree on a common marking system. Care should be taken in watering as a heavy stream of water could dislodge seeds.

FIGURE NO. 7 CORN SEED ORIENTATIONS

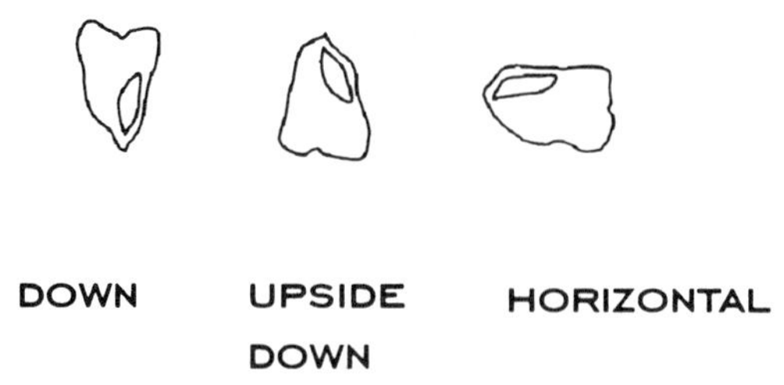

4. Explain and demonstrate the use of the protractor for measuring angles. This was a difficult concept for our students and may be aided with additional exercises on the use of a protractor. It is easiest to measure the angle of root and shoot growth with respect to horizontal. Therefore, a root growing vertically will have a measured angle of -90°.

5. After about five days, students will dig up one "outside row." Two or three days later, the second outside row of seeds will be unearthed. The last row will be dug up in two to three more days. Students will observe germination and measure root lengths and the angles formed with respect to horizontal during germination. These results should be noted in the lab book along with the results from other groups (i.e., other orientations).

6. Students will organize data and draw conclusions. This goes in the lab book and can serve as the basis for a windup discussion.

NOTEBOOKS

Blackboard Information

Gravitropism DATE: ____________________

OBJECTIVE: To test the hypothesis that gravity influences the direction of root (radicle) and shoot growth.

PROCEDURES:

1. Select 18 corn seeds.
2. Fill the potting container 1/2 full with soil.
3. Carefully sow seeds in the orientation assigned to your group, three rows of six seeds. Mark the rows clearly and label the box.
4. Cover the seeds with 3 cm of soil and moisten the soil gently. Do not dislodge the seeds.
5. On (designated day), carefully dig up one "outside" row of seeds and measure root and shoot angles with respect to horizontal. Share your results with groups who used the other orientations. Record results in your lab book.
6. On (designated day), dig up the second "outside" row of seeds and do as before.
7. On (designated day) dig up the last row and do as before.

Results

The data from each group need to be put in a table (see next page) on the blackboard and then copied into each student's notebook.

Root Length and Orientation					
	Variable	Date 1	Date 2	Date 3	Comments
Down:					
Shoot Length					
Shoot Orientation					
Root Length					
Root Orientation					
Upside Down:					
Shoot Length					
Shoot Orientation					
Root Length					
Root Orientation					
Horizontal:					
Shoot Length					
Shoot Orientation					
Root Length					
Root Orientation					

Discussion Ideas

1. What did we learn?

2. Regulation is a major life function. How is it illustrated in this experiment?

3. What role would hormones play in the gravitropism response?

4. Should seed orientation be a major concern to a farmer when he or she is planting a crop?

5. What might happen to root and shoot growth in weightless conditions?

6. What is the current explanation for the mechanism that makes the shoot grow up and the radicle grow down?

7. How might a space station create conditions of gravity for germinating seeds?

Further Activities

1. Germinate seeds in an alternate gravity field. The turntable of an old phonograph could be used to establish an additional gravity field in which to germinate seeds.

2. Investigate participation in some of the projects on microgravity sponsored by NASA on various shuttle missions.

PHOTOTROPISM 9

BACKGROUND

Educational Opportunities for Students

A. Learn of the influence of light on plant development.
B. Gain additional experience in the use of a protractor to measure angles of plant growth.

Biology Background

Leaves appear green because the pigment chlorophyll is especially effective in absorbing light in the blue and red wavelengths and less effective in the green wavelengths. (Light, chlorophyll, and photosynthesis are explored in separate experiments.) Therefore, little light at blue or red wavelengths is reflected or transmitted by leaves. It is this fact that allows developing plants to react to competition for light from neighboring plants.

Most plant tissues have a pigment called *phytochrome* that absorbs red wavelengths. In elongating cells in which phytochrome has absorbed red wavelengths, the pigment triggers a hormonal sequence of processes that causes cells to elongate to their "normal," relatively compact length. However, if phytochrome in a cell fails to absorb sufficient red light, then the cells continue elongation and become very long. Developing seedlings under dense layers of leaves, such as in a forest, will have long, spindly stems because the ared wavelengths are absorbed by the leaves of the older plants above the seedling causing phytochrome in the seedling to signal for elongated cell development.

Phototropism is a special consequence of the control of elongated cell growth signaled by phytochrome in the absence of red light. Cells on the "dark" side of an elongating stem are receiving less red light than the cells on the "bright" side of the stem. As a result, the cells on the "dark" side of the stem elongate more than the cells on the "bright" side, and the stem appears to grow towards the light. (It is

FIGURE NO. 8 PHOTOTROPISM RESPONSE

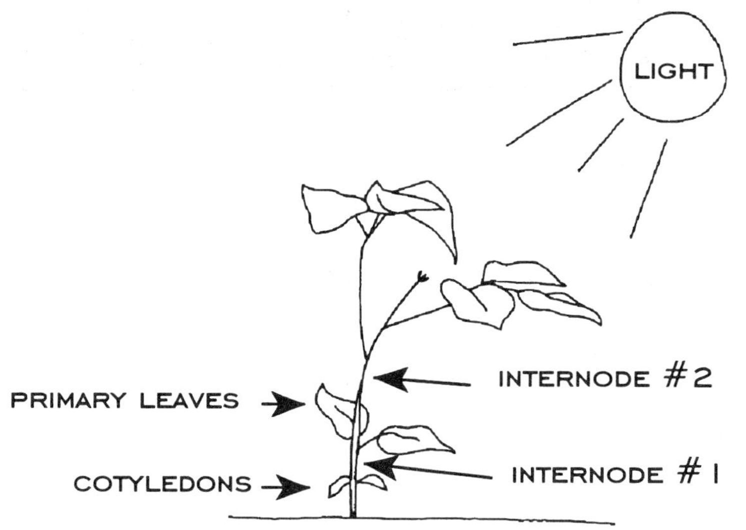

probably more appropriate to describe the stem growth as being growth away from the dark rather than growth towards light.) The advantage of phototropic growth is that a developing seedling will position its leaves so that the exposure of leaves to light for photosynthesis is maximized. ("Shade" type plants that have adapted to growth in low light conditions usually have no or only a weak phototropic response.)

One of the treatments in the experiment uses blue plastic wrap placed on windows. Although this wrap looks blue to the human eye, some red light is transmitted by the plastic wrap. Therefore, a phototropic response is also likely to be observed in this treatment by students. However, it may be that measurements of lengths and angles will show less of a response than the plants exposed to white sunlight. Of course, those seedlings in the dark or shaded treatment should exhibit the greatest stem lengths.

Materials and Experimental Environment

- Bean seeds
- Recycled styrofoam cups (three for each lab group)
- Potting soil
- Blue plastic wrap
- Stakes
- Metric rulers

Materials and Experimental Environment continued

- Protractors
- Shelf area for three light treatments

Procedures

1. Each lab group will sow three bean seeds each in three styrofoam pots.

2. Cups should be labeled as usual, then a vertical line drawn on the side. This line will function as an orientation cue so that students can always replace the cups in the same position relative to the light source.

3. Each group will place a pot in each of three environments:

    a. sunlight (control)
    b. blue light (We covered some window panes with blue plastic wrap.)
    c. partial dark (We placed the pots in book shelves under those on which pots of treatment a and b were sitting.)

4. Students should tend their cups and watch for germination, thinning to one seedling when the plants are approximately 8-cm tall. Plants should be ready for making measurements at this stage.

5. To make observations, students will measure the length and angle of internodes of the plant at three locations:

    a. from cotyledons to primary leaves
    b. from primary leaves to first trifoliate leaf
    c. from first trifoliate leaf to second trifoliate leaf.

6. Observations are made every two or three days until stem sections being observed stop elongating. Stakes may be needed to provide support for the plants.

7. Graphs should be encouraged for the notebooks as a way to visualize the differences in the development of length and angle.

NOTEBOOKS

Blackboard Information

Phototropism DATE: ____________________

OBJECTIVE: To observe phototropic responses in the development of bean seedlings.

PROCEDURES:

1. Prepare three styrofoam pots with soil.

2. Sow three bean seeds in each cup (about 2-cm deep).

3. Label the cup, then draw a vertical line on the side. This will help you replace the cup in the same position after you make an observation.

4. Place one pot in each of three treatments: a. control (in sunlight), b. blue light, c. partial dark.

5. Check your pots daily and keep the soil moist. When the seeds have germinated and you have plants about 8-cm tall, thin to one seedling per cup.

6. Begin measurements on length and angle with respect to horizontal for three stem segments: cotyledon to primary leaves, primary leaves to first trifoliate leaf, first to second trifoliate leaf.

7. Check seedlings every two or three days, making and recording measurements.

Results

Make one of the following tables for each treatment (see next page).

Treatment________________________________

Date	Cotyledons/primary leaves		Primary leaves/1st trifoliate		1st trifoliate/2nd trifoliate	
	length	angle	length	angle	length	angle

Discussion Ideas

1. What did we learn?

2. What makes leaves appear green?

3. What does the pigment phytochrome do for plants?

4. What happens when phytochrome doesn't get enough red light?

5. What would you expect seedlings shaded by many large trees to look like?

6. Why is it more precise to say that plants grow away from light?

7. How do we account for plants that "like" shade?

Further Activities

1. Grow plants near or under other plants to observe growth pattern of shaded plants. If there is garden space, various sowing patterns could result in different degrees and patterns of shading.

2. Test other plant species for phototropic response. Selection of shade tolerant species would give interesting comparisons.

3. Obtain other color filters and observe plant growth. There are several colors of plastic wrap.

RESPIRATION 10

BACKGROUND

Educational Opportunities for Students

A. Gain experience doing simple chemical assay.
B. Learn of differences in biochemical composition of tissue.
C. Learn importance of respiration in living tissue.

Biology Background

Respiration is the biochemical process where stable, chemical energy stored in various compounds is transferred to specialized molecules (for example, ATP and NADH) which are readily used to "fuel" a number of processes. Aerobic respiration, which involves the consumption of oxygen, is most common among higher organisms. Assuming that the original storage of energy is a carbohydrate [$(CH_2O)_n$] and that the product of respiration is ATP, then respiration can be written in its simplest equation form (n = 1) as,

$$(CH_2O) + O_2 + ADP \longrightarrow CO_2 + H_2O + ATP.$$

In addition to carbohydrate, a very efficient storage of chemical energy is in lipids or fats. Lipids store much more energy per unit weight than carbohydrate. It is no accident that mammals, including people, have evolved systems to store excess chemical energy as fat. Therefore, for those tissues with a high lipid content, biochemical reactions are needed to break down the lipid into the carbohydrate formulation. In its simplest form this process can be written in an equation as,

$$\text{lipid} + O_2 + ADP \longrightarrow (CH_2O)_n + ATP.$$

In this experiment, the students have the opportunity to examine the importance of oxygen in allowing respiration and seedling growth to be sustained. By restricting the available oxygen supply in a sealed container, the amount of seedling growth is restricted.

Those germinating seedlings in smaller containers will have less respiration, and therefore less growth.

Because the storage of carbohydrate in comparison to lipid results in very different levels of energy storage, the initial exercise for the students in this experiment is to compare the composition of a corn seed (high in carbohydrate) versus a peanut seed (high in lipid). After testing the seeds of these two species for biochemical composition, the students should be able to hypothesize on possible differences in oxygen consumption and seedling growth by the two species in a sealed atmosphere.

Materials and Experimental Environment

- Bean, corn, and peanut seeds
- Clear glass jars of varying volumes with airtight lids
- Graduated cylinders
- Metric rulers
- Single-edged razor blades
- Paper towels
- 20% iodine solution in dropper bottles
- Dark shelf or cabinet area to store respiration chambers

Procedures

1. Ask students several weeks before this exercise to bring in glass jars. The best jars are clear and have wide mouths. There should be a set of five or six jars of decreasing volumes for each class period. The largest can have a volume of approximately one liter and the smallest a volume of about 100 mL. The importance of airtight lids cannot be overemphasized, as it is critical to the experiment.

2. Organize lab groups so that there is one jar for each group. A range of volumes should be used in each class. Groups determine the volume of their jars by filling them with water and measuring the water volume. The initial volume of oxygen in the jar is calculated by knowing air is 21% oxygen. Before putting the formula on the board, discuss possible ways to determine these volumes. Students are likely to have the knowledge to arrive at the correct procedures.

3. Students will prepare the experiment by folding two paper towels, wetting them thoroughly, and placing them in the

bottom of their jars. Twelve bean seeds will then be placed on the wet towels. The jars will then be closed tightly, labeled, and placed in a dark location, such as a cabinet.

4. After four or five days, the students can remove the jars from the dark location long enough to estimate shoot length (the lid must remain tightly closed so that more oxygen will not be introduced). The jars will then be replaced in the dark cabinet.

5. In another three or four days, the jars may be opened and the shoot lengths measured. Students now have enough data to construct a graph plotting average length of shoots against oxygen volume. Students will probably raise the question about how to handle ungerminated seeds when calculating averages. Encourage debate. You may wish to suggest that students plot the data obtained both by including and excluding the ungerminated seeds in calculating an average.

6. Graphing of the results of this experiment will be a challenge for students because time is not the significant variable of this exercise. The students should think of oxygen volume as the important independent variable, plotting it on the x axis.

7. On a separate day, the students can investigate variation in the chemical compositions of corn and peanut seeds. Each student will test a corn seed and a peanut seed for starch and lipid composition. The seeds are split in half (for the corn seed, place it embryo side up and cut longitudinally). Lipid in the seed is tested by rubbing one half of each seed on paper toweling. The high oil content of a peanut will leave an oil spot on the paper towel. An iodine solution is used to test for starch content.

This is the first experience with a chemical solution in this series of exercises. While there is little concern in using an iodine solution, general laboratory safety in handling chemicals should be reviewed with students. Iodine is used in this test because it will turn starch black. The same reaction will also happen between iodine and cellulose, so students should be warned to avoid getting the solution on their clothes because it will stain. If laboratory aprons or coats are available they should be worn for this exercise.

The presence of starch in the seeds is tested by dipping the remaining seed halves in iodine solution or putting a drop of solution on the cut surface of the seeds. The solution only needs to be left on the cut seed surface for about 10 seconds, and then it can be washed off with tap water. The iodine will turn the endosperm of the corn seed a dark color because the endosperm is mainly starch. The cellulose in the cell walls of the peanut seed will result in only a light gray color. Students will record relative amounts of starch and lipid on a chart.

NOTEBOOKS

Blackboard Information

Respiration DATE: ____________________________

OBJECTIVE: To test O_2 requirement for respiration.

$(CH_2O)_n + O_2 + ADP \longrightarrow CO_2 + H_2O + ATP$

$lipid + O_2 + ADP \longrightarrow (CH_2O)_n + ATP$

PROCEDURES:

1. Each group selects a glass jar and determines O_2 volume (Hints: O_2:air = 0.21 and 1 mL = 1 cm^3).

2. Fold two paper towels together, wet thoroughly, and put in jar.

3. Put 12 soybean seeds on the towel.

4. Label and close jar; put in dark location.

5. Estimate shoot length on __(date)__; make final measure of lengths on __(date)__.

Results

Data from each group are to be recorded in a table on the blackboard so that each student can put all data into his/her notebooks.

		Average Shoot Lengths		
Jar	O_2 volume	(Date) (est.)	(Date)	Comments
I				
II				
III				
IV				

Students should construct a line graph using the data from all groups. That is, average shoot lengths should be plotted against O_2 volume.

Seed Composition DATE: ____________________

OBJECTIVE: To test seeds for different biochemical compositions.

$(CH_2O)_n + O_2 + ADP \longrightarrow CO_2 + H_2O + ATP$

$\text{lipid} + O_2 + ADP \longrightarrow (CH_2O)_n + ATP$

PROCEDURES:

1. Each person gets one seed of peanut and one seed of corn.
2. Split seeds in half.
3. Dip one seed half in 10% iodine.
4. Rub the other seed half on a paper towel.
5. Note and record relative amounts of $(CH_2O)_n$ and lipid in each seed species.

Results

Relative Compositions of Starch and Lipid			
Treatment	Corn	Peanut	Comments
paper towel			
iodine			

Discussion Ideas

1. What have we learned?

2. What is the value to animals of storing fat?

3. Which would be better for you to eat — corn or peanut? Why?

4. How do you think corn and peanut seeds would compare in O_2 consumption if they were placed in the different jars? Explain.

Further Activities

1. Compare amount of growth in respiration jars between seeds with high and low lipid contents. Soybean and peanut are two species with high lipid content. Compare similar weight seeds (e.g., bean seeds) for similarities. Determine final seedling weight and size of each species.

2. Design and construct a sealed system so that oxygen consumption in respiration is matched by oxygen production from photosynthesis.

3. Examine respiration in the absence of oxygen by setting up a fermentation experiment using, for example, apple cider.

CHLOROPHYLL 11

BACKGROUND

Educational Opportunities for Students

A. Expose students to advantages and disadvantages of cooperative investigation by a large group.
B. Learn of approximate time frame required for new biochemical syntheses.
C. Learn of function of chlorophyll in plants.

Biology Background

Before discussing chlorophyll, students need a brief introduction on light and colors. Students should be reminded that "white" light is composed of a spectrum of all colors. The color of an object is perceived by the human eye because that object did not absorb particular wavelengths. (This is a difficult concept for some students.)

Chlorophyll is the most abundant pigment on earth and is the primary molecule for "trapping" the solar energy that reaches plant tissue. The structure of chlorophyll makes it particularly effective in absorbing the red and blue wavelengths. Although chlorophyll also absorbs a substantial amount of light in the green wavelengths, it is less effective in absorbing these wavelengths; and hence the human eye "sees" chlorophyll as a green pigment.

Chlorophyll is efficient in passing the energy trapped from light to other molecules so that other molecules may form high energy chemical bonds. It is the use of light energy to form chemical energy that is the basis for most plant life, and subsequently nearly all life forms on earth. The advent of chlorophyll was truly a momentous event in evolution on earth.

In this experiment the cooperation of all classes in the school day will be required for several days to document the greening of seedlings. The development of the green coloration is, of course, a consequence of the synthesis of chlorophyll. Increases in the intensity of green coloration are indicative of the amount of chlorophyll.

Chlorophyll is synthesized only in tissue after being exposed to light. Light triggers a biochemical signal to initiate the production of chlorophyll. The seedlings in this experiment should become fairly green after exposure to light in about 24 to 48 hours under moderate temperatures. This time frame for synthesis is roughly the time frame required to initiate and synthesize in quantity many biochemical materials. Students might be able to relate this 24- to 48-hour time frame to the length of their last flu and the time it took for their bodies to synthesize the materials to overcome the flu infection.

Materials and Experimental Environment

Exercise needs to be started early in the first week so that the seedlings are at the proper stage for observation in the second week.

- Bean seeds
- Germination chambers
- Paper towels
- Masking tape
- Plastic wrap
- Green color strips from a paint store are useful as color references.
- Dark, warm place for germination chambers
- Sunny locations to expose seedlings to light

Procedures

1. Give students an introduction to light and colors, emphasizing that to our eyes, color is what is not absorbed by a material.

2. Each lab group prepares a germination chamber and puts 12 bean seeds in it (three paper towels, water). The chambers should be completely sealed with masking tape so that they are light tight. The actual experiment will be in the latter part of the following week so the germination needs to be started on a Monday to Wednesday.

3. The seeds will be allowed to germinate for 7 to 10 days in a

warm environment. (If this experiment takes place in cold weather, it may be necessary to move the germination chambers to an area in the school that will be constantly heated.)

4. After 7 to 10 days, the trays are opened and exposed to light. The paper towel covering the seedlings should be discarded. The top tray of the germination chamber should be replaced with clear plastic wrap to avoid drying of the seedlings left in the bottom tray.

5. Students will observe and record the color of the seedlings over three days. Therefore, the opening of germination chambers should be no later in the week than Wednesday. The experiment is complicated because each class opens its seedling trays at different times during the day. Each class needs to make notes for its trays and for all trays that have been previously opened in other classes. A lesson in organizing large experiments will be taught here. Put a chart on the board (see blackboard information) and record results for other classes to copy.

6. Coloration estimates for this experiment are somewhat subjective, depending on each individual's perception of color and shade. The students will have to arrive at a descriptive system to use. A gradation of colors from light yellow to deep green needs to be described. Color strips obtained from a paint store make a useful reference.

NOTEBOOKS

Blackboard Information

Chlorophyll DATE: ____________________

OBJECTIVE: To study the speed of chlorophyll synthesis.

PROCEDURES:

1. Prepare a germination chamber with paper towels and water and put 12 bean seeds in it.

2. Seal the chamber with masking tape to make it light tight and leave it in a warm place for 7 to 10 days.

3. After 7 to 10 days, remove the top from your germination chamber, discard top paper towel, and cover bottom of the chamber with plastic wrap to keep seedlings moist.

4. Place tray with seedlings in a location exposed to light.

5. Record the color of your seedlings and those of the other classes for the next three days. Include observations of light conditions (sunny, partly cloudy, etc.).

Results

The observations from each class need to be entered into the following table. Each student should copy the table into his/her notebook. [NOTE: This table needs to be developed for each of the three days of observation.]

Date______________	Observation Time			
Class where seedlings originated	10:55	11:35	1:00	1:45
3rd				
4th				
5th				
6th				
7th				
Light conditions				

<u>KEY:</u>

W	=	white
Y	=	yellow
G	=	green
PG	=	pale green
W-YG	=	white to yellowish green

Discussion Ideas

1. What did we learn?
2. How long did it take for the seedlings to become green?
3. Did the amount of light seem to matter?
4. What do we see when we look at white light?
5. When we see a particular color, like red, what are we actually seeing?
6. What job does chlorophyll do in the plant?
7. What makes chlorophyll appear?
8. How does intensity of green relate to amount of chlorophyll?
9. How does the length of time required for chlorophyll match our biological syntheses?

Further Activities

1. Perform the coloration experiment in reverse. That is, put green plants in darkness and record loss of green color with time.

2. Explore the factors associated with loss of chlorophyll in many leaves in the autumn.

3. The perception of color by the human eye can be tested by using combinations of differing filters. (Plastic wraps of differing colors could be used.)

STOMATA & PHOTOSYNTHESIS 12

BACKGROUND

Educational Opportunities for Students

A. Gain experience with microscopic observations.
B. Learn about some of the factors associated with photosynthesis.

Biology Background

Photosynthesis occurs in two stages. The first is the photochemical capture of light energy and its temporary storage in high energy chemical bonds. (The chlorophyll experiment exposed students to this first step in photosynthesis.) The second stage — the one of primary focus of discussion in this experiment — uses the energy temporarily stored in the high energy bonds of ATP and NADPH to accumulate organic compounds in the plant.

The simplest biochemical expression of the synthesis of organic compounds from carbon dioxide (CO_2) by photosynthesis is given by the following equation.

$$\text{Light} + CO_2 + H_2O \longrightarrow (CH_2O)_n + O_2.$$

Of course, in the absence of light there is no energy source for photosynthesis and there is no assimilation of carbon dioxide. The failure to find sugars and starch in the absence of light is the basis of this experiment.

Uptake of CO_2 by leaves presents a special problem to land plants. Because carbon dioxide is available in the atmosphere around plants and photosynthesis takes place in the cells of leaves, CO_2 must move from the atmosphere to inside leaves. To admit CO_2 into leaves, leaves have small pores in their surfaces called *stomata.* The stomata are formed by a pair of *guard cells* that function to vary the width of the pore. In light the guard cells swell, causing the

FIGURE NO. 9 DICOT STOMATA

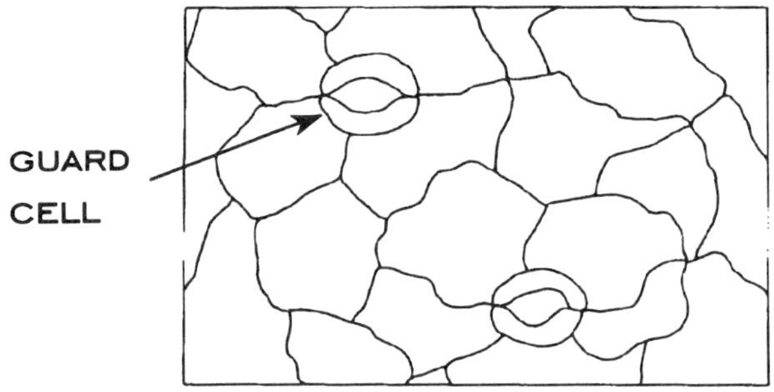

pore to be at its widest, and CO_2 diffuses into the leaf and into the cells to be assimilated in photosynthesis. In the dark or under drought conditions the guard cells are not turgid, the stomata are closed, and there is no photosynthesis. Opening of the stomata not only allows CO_2 to diffuse into the leaf, but allows water vapor to diffuse out of the leaf. (This important process of water loss by land plants is considered in the transpiration experiment.)

The organic product of photosynthesis is given as $(CH_2O)_n$ which is the basic component of carbohydrate. In putting six of these units together (that is, n=6), the simple sugars (usually glucose or fructose) are produced. Combining glucose and fructose gives sucrose (n=12) which is the most abundant sugar found in nature. Sucrose is the ordinary table sugar that is used in soft drinks and candy bars. Depending on the nature of the bonding between the simple sugars, long strings of simple sugars result in starch and cellulose. Students can be reminded that thanks to photosynthesis they have french fries to eat (starch), cotton clothes to wear (cellulose), and wood pencils and paper for doing their homework (cellulose)!

The other important product of photosynthesis is oxygen. Of course, without the evolution of oxygen into the atmosphere as a by-product of photosynthesis in the earlier stages of earth's evolution, most animal life could not have evolved. The generation of oxygen by photosynthesis in water plants can be shown in an additional experiment where gas bubbles are formed in the water.

Materials and Experimental Environment

- *Setcreasea* leaves (or any others with easily observed stomata and guard cells)
- *Coleus* plants (white and green variegated leaves — use those started in the asexual reproduction experiment, if possible). At least two plants will be needed.
- Single-edged razor blades
- Two petri dishes per class
- Microscope and slides
- Forceps
- 20% iodine solution
- Large cardboard box to cover plants
- Sunny location to expose plants to light

Procedures

1. Discuss the various features of photosynthesis. (This will depend on the level of students you teach.)

2. Students will examine leaf surfaces of *Setcreasea* under the microscope. They should locate, draw, and label a stoma with guard cells in their lab books.

3. Students will observe the pigmentation of the *Coleus* plant, and will place one plant in a sunny location and the other in the dark. A heavy cardboard box with a cover can serve as a dark chamber. Plants should be checked daily and watered when necessary.

4. After subjecting the plants to about five to seven days of the two light treatments, harvest leaves for the iodine test. (NOTE: one leaf can accommodate more than one lab group.) The teacher should cut 2- to 3-mm wide strips from leaves grown under both conditions. The strips should be cut to include both green and white areas. To avoid confusion, have students cut a small notch in all strips at the green end.

5. For each class, label petri dishes according to light treatment, add iodine solution, and have students leave their strips in the solution overnight. The iodine solution can be made by diluting a small bottle of iodine purchased at the drug store. A dilution of 20% iodine in water works satisfactorily.

6. After 24 hours, lab groups will remove a leaf strip from each treatment and examine it under the microscope. The strips should be drained briefly on a paper towel before being placed on a microscope slide. Forceps are the best way to handle the small strips.

7. When students check the strips under the microscope they should check each pigment area. They should look for the characteristic dark starch granules and record their observations. The iodine will generally enter the leaf strips from the cut edges. Therefore, the students may have the most success in locating starch granules at the cut edge of the leaf strips.

NOTEBOOKS

Blackboard Information

Stomata and Photosynthesis DATE: ___________________

OBJECTIVE: To identify stomata and relate them to the process of photosynthesis, and to show that starch (and subsequently all organic matter) is the product of photosynthesis.

PROCEDURES:

1. Using the microscope, look at the underside of the *Setcreasea* leaves. (Fold the leaf with the upper leaf surface on the outside, then tear away a piece of the epidermis on the underside and place it on a slide. Add a drop of water.)

2. Locate, sketch, and label a stoma with its guard cells.

3. Observe the pigmentation of a *Coleus* leaf.

4. With the class, place a *Coleus* plant in a sunny spot and another in a place where it will receive no light. Keep the plants watered.

5. After about one week examine the leaves for starch.

   a. Get a 2-3 mm wide leaf strip from each treatment. Cut a notch in the green end of the strip. (This will help later to identify each pigment area.)

   b. Soak the strips overnight in petri dishes containing iodine.

   c. Next day remove a strip from each treatment, drain excess iodine off on a paper towel, and then examine them under the microscope for the stain pattern created when iodine and starch combine. Starch granules will be stained a dark color. Record results.

Discussion Ideas

1. What did we learn?

2. There are two stages of photosynthesis. What happens in each?

3. Give a simple equation to express what happens in photosynthesis.

4. What happens when the plant does not receive light?

5. How do guard cells work?

6. What do french fries, cotton, and pencils and paper have in common?

7. Besides sugar, what other substance is a product of photosynthesis? Why is this important?

Further Activities

1. Examine different plant tissues for the presence of starch when subjected to various treatments.

2. Obtain *Elodea* (available in most stores with aquarium items) and observe generation of air bubbles (oxygen) under differing conditions.

3. Examine amount of photosynthesis under different light conditions by growing plants, harvesting at weekly intervals, and measuring their dry weight. This is probably best done as a summer project when a large number of plants can be grown in the garden for this experiment. Shading by other plants or shading cloths can be used.

TRANSPIRATION 13

BACKGROUND

Educational Opportunities for Students

A. Provide more experience in measuring metric weights.
B. Learn about diffusion processes.
C. Learn about the high water requirements in transpiration by plants.

Biology Background

Diffusion of molecules or particles in a fluid is dependent on the concentration gradient of the diffusion material and on the resistance to diffusion. Transpiration is the diffusion of water vapor molecules from inside leaves through other air molecules (which is the fluid) to the atmosphere outside the leaves. The concentration gradient is the difference between the water vapor concentration of the air inside the leaf and the concentration of the water vapor in the atmosphere outside the leaf. The main resistance to diffusion in this pathway results from the stomata pores through which the water molecules must diffuse.

As discussed in the stomata experiment, deformation of the guard cells determines the width of the stomata pore and, consequently, the resistance of the stomata to water vapor diffusion. Generally, stomata are open in the light, and the loss of water is large. In the dark, where rates of photosynthesis are low, the stomata width is small and water losses are low. In this experiment, differences in stomata resistance are achieved by placing one plant in the light and another in shaded conditions.

The concentration gradient of water vapor is also manipulated in this experiment by placing plants in "humidity chambers." These chambers retain the transpired water and the humidity becomes substantially greater than the humidity of the normal room atmosphere. The increased humidity around these plants should result in a lower transpiration rate.

FIGURE NO. 10 HUMIDITY CHAMBER

Because the water vapor concentration gradient between the inside of leaves and the atmosphere is usually very large, the amount of water required by plants is large. Of course, those plants growing in hot, arid regions require much more water than those plants in a tropical rainforest. Plants in hot, arid regions generally have evolved specialized stomata that cause stomatal resistance to water vapor diffusion to be very high. Plant canopies growing in temperate regions during the summer have water loss rates of about 4 to 8 mm (mm^3 of water per mm^2 of ground area). To avoid serious drying of the soil during the summer in these temperate regions, lawns, crops, and forests all need on the average a 25-mm rain about every three to six days.

Materials and Experimental Environment

- Styrofoam cup pots (recycle)
- Bean plants (The class can sow the seeds in styrofoam cups about two weeks before this exercise)

Materials and Experimental Environment continued

- Two to four humidity chambers per class (2-liter clear plastic soda bottles with the base portion removed, and the cap left on)
- Balance scales and gram masses
- Graduated cylinders
- Sunny and shaded locations to place plants

Procedures

1. About 10 days before doing this exercise, have students sow three seeds in each of the styrofoam cup pots. Thin to one plant in each pot after plants are about 8-cm tall.

2. The day before the experiment is to begin, have the students water each pot abundantly.

3. Divide students into four groups. Each group will use one or two seedlings and be responsible for one of the treatments:

    a. light - Pots will be labeled "light" and placed in a sunny location.
    b. light and humid - Pots will be labeled "light and humid." A humidity chamber is placed over these plants. Plants are put in the sunny location.
    c. shade - Pots will be labeled "shade" and placed in a shady location.
    d. shade and humid - Pots will be labeled "shade and humid." A humidity chamber is placed over these plants. Plants are put in the shady location.

4. Students will label and weigh each plant and record the results. The plants should then be placed in the treatment area assigned to the group.

5. Every two or three days the students will weigh and record pot weights. They should then calculate the water loss from each pot.

6. The students will then add exactly the same amount of water lost back into the pot. If the students do not know the relationship between grams and milliliters (one ml of water

weighs approximately one gram), this is a good way to have them learn this important concept. Repeat the water loss cycle about four times.

7. The data obtained in this exercise can be analyzed in a number of ways. Calculating the ratio of two treatments should indicate the differences in transpiration. For example, each set of data could be explored by calculating the ratio of each of the other three treatments against the "light" treatment.

NOTEBOOKS

Blackboard Information

Transpiration DATE: ___________________

OBJECTIVE: To examine the amount of water required by plants.

PROCEDURES:

1. Abundantly water pots on the day previous to the start of the experiment.

2. Weigh pots, label, and place in treatment area.

3. Four investigation groups will each do one treatment:

    a. light (control)

    b. shade

    c. light and humidity chamber

    d. shade and humidity chamber.

4. Every two or three days, weigh, record, and calculate water loss. (Remember that 1.0 g of water = 1.0 ml of water.)

5. Rewater each pot with the amount of water that had been lost.

Results

The results obtained by each group should be entered in the following table (see next page) on the blackboard. Each student should copy the table into his/her notebook.

Observation	Light	Light & Humid	Shade	Shade & Humid	Comments
Initial weight					
weight					
H_2O loss					
weight					
H_2O loss					
weight					
H_2O loss					
weight					
H_2O loss					

Discussion Ideas

1. What did we learn?
2. By what process does water move out of a plant?
3. What degree of opening are the stomata when in the light? In the dark?
4. Why would a humidity chamber keep transpiration rate low?
5. Why do plants lose so much water by transpiration?
6. How much water would be needed in a week for 1,000 plants similar to ones placed in the light?

Further Activities

1. Compare the water loss rate of different size plants.
2. Take a section of grass sod and put in pot to estimate the amount of water used by lawns on each day (mm^3 water lost per mm^2 of the surface area of the pot). Placement of the pot on the lawn is important; shade vs. sunny location, near bushes or center of lawn, etc.
3. Look at stomata structure of various species, including cactus, to see if these differences may be important in controlling plant water loss.

COMPETITION 14

BACKGROUND

Educational Opportunities for Students

A. Apply observational skills and knowledge to observe plants growing in competition.
B. Introduction to competition and interaction of natural ecosystems.

Biology Background

Any organism in its natural state is always interacting with both its physical environment (temperature, light, water, etc.) and other living organisms. The interaction with other organisms can range from predatory activity to a competition for the same resources. In ecosystems where water and temperature are adequate, light can be the resource for which plants will compete. The tropical rainforest is an example where this competition among plant species is particularly important.

In this experiment, two species are sown in a small area and forced into a situation where light availability is important. A fast-emerging, low-growing species (radish) will initially flourish but should be overtaken by a slower growing but taller species (squash). Observations and interpretation of this experiment will draw on students' previous experience with chlorophyll and light, photosynthesis, and phototropism. The accidental introduction of insects or diseases into the experiment may provide an additional competitive factor.

The competition experiment can be extended to a field trip where students are asked to make observations on the important competitive factors they may find in a small "ecosystem." A lawn, flower bed, vegetable garden, or a small wooded area is all that's required for a challenging ecosystem for study. The students can draw on their various experiences from the hands-on exercises to think about the type of plant species in the ecosystem, the growth pattern of the various plant species, and assuming this is done in the spring, they may be able to speculate about their reproductive patterns. Students should be encouraged to look for non-plant organisms as important components of the ecosystem and potential predators of plants. By this time the students should know they need to get their hands dirty and also look at belowground features!

Materials and Experimental Environment

- Radish and squash seeds
- Rectangular containers, 7-cm or more deep (The plastic fruit/vegetables containers used in the gravitropism experiment work well.)
- Potting soil
- Sunny location for growing plants

Procedures

1. Discuss competition and let students speculate about what might be important competitive factors.

2. Organize students into lab groups and discuss the design of the experiment. Each group will fill a container with potting soil and sow both radish and squash seeds in high densities. Water will be added as usual and containers placed in a sunny spot.

3. Set up an observation schedule for the class (about every three to four days is good). Students will look at their containers, tend them, and record what they see. Encourage students to give thought to the descriptive terms they choose for their comments.

4. The teacher will decide when the experiment is at an end. There are many things to learn at each stage. Some groups may neglect to water their container and thus alter the competition.

5. Lab groups will organize data, draw conclusions, and record them in their lab books.

NOTEBOOKS

Blackboard Information

Competition DATE: ____________________

OBJECTIVE: To observe the growth of two plant species in competition.

PROCEDURES:

1. Fill plastic container with soil.

2. Sow approximately 12 squash seeds 3-cm deep and 36 radish seeds 1-cm deep. Water carefully and place in a sunny location.

3. Observe each species every three or four days. Include information on plant height, number of leaves, and size of leaves.

Results

Each student should develop the following table:

Date	Observations				Comments
	Radish		Squash		
	height	leaves	height	leaves	

Discussion Ideas

1. What have we learned?

2. What is competition?

3. For what might plants compete?

4. What are some characteristics plants have evolved to make them better competitors?

Further Activities

1. Design experiments to test competition for other factors such as water or minerals.

2. Compare ecosystems (e.g., desert vs. rainforest) and describe the crucial factors that influence competition among plants. How might animals influence plants and plant competition?

3. Study in detail a small area of a readily available ecosystem. Document what plants exist in the ecosystem and their abundance. What factors influence plant growth traits and plant populations? How does the ecosystem change through the year as competitive factors change?